CONCEPTUAL UNDERSTRUCTURE OF HUMAN EXPERIENCE

VOLUME 2 • ARTWORKS

WES RAYKOWSKI, Ph.D

Copyright

Copyright © 2014 by Wes Raykowski, Ph.D

All rights reserved under Pan American and International Copyright convention

Contact details

conceptualunderstructure@gmail.com

ISBN: 1494852608
ISBN 13: 9781494852603
Library of Congress Control Number: 2013923836
CreateSpace Independent Publishing Platform
North Charleston, South Carolina

PREFACE

This volume is a companion book to *"Conceptual Understructure of Human Experience: Thesis"*. It presents the artworks and discusses their role in the research leading to the doctoral thesis.[1]

While studying fine art, I became intrigued by the consistent use of diagonal lines in expressions of movement; of the vertical orientation for conveying various notions of vitality; and of the horizontal orientation for death and general absence of energy. On the basis of subsequent investigation of expressions in fields other than fine art, I argue that the two canonical orientations are part of a more general concept (described in the thesis as *experiential schema*) which connects *intensity* (or *value*) with its *extent* – where values represent the properties of physical objects interpreted as units, and intensity the sensory qualities referring to these objects, and where extent stands for the repetition of the unit. This suggests the presence of at least two scales in human conceptualisations: the smaller scale of the unit and the larger scale of the collection of units combined together into a single expression such as *two important issues* or *five tasty apples*. The details of the experiential schema could be found in the book *"Conceptual Understructure of Human Experience: Thesis"* which has to be purchased separately.

The present volume includes some of the artworks created in the course of doctoral research. The works are organised thematically to demonstrate various aspects of the thesis: the notion of space and content, the issue of unit indivisibility and emergence of scales, the idea of levels in the context of comparison, the role of height and width, lines and their orientation, and a few ways of creating contrasts. A brief discussion about art making as a research methodology concludes this part of the book. Appendix B includes a short article about one of the artworks displayed at the **Collaberatum: UQCA @metr** exhibition organised by The University of Queensland, Metro Arts, and Griffith University Queensland College of Art.

---

[1] The content of this volume was located in Appendix A and B of the original submission.

INDEX

Preface. iii
Index · v

APPENDIX A 1
1   INTRODUCTION 1
2   INVAGINATION. 3
3   CONTENT 9
4   DEFINING SPACE. 11
5   INVESTIGATING HEIGHT AND WIDTH 14
6   LEVELS, HEIGHT AND COMPARISON 16
7   EMERGENCE OF LARGER SCALE. 19
8   INDIVISIBILITY OF UNITS. 22
9   CONTRASTS 25
10 LINES AND THEIR ORIENTATION. 29
11 CAN ART MAKING BE A RESEARCH. 36

APPENDIX B 39

LIST OF FIGURES RELATED TO ARTWORK CREATED BY THE AUTHOR 42

APPENDIX A

1 INTRODUCTION

Even before my involvement with fine art, I was captivated by the issue of life and death. This preoccupation was reflected in my first independent works: *House at Nobby Beach* and *1m Long Construction Timber*.

Fig 1: Raykowski W, 2002, *House at Nobby Beach*, sand and sea water, 70cm x 35cm x 70cm; nonextant

The image above shows two forms – the void on the right hand side was made by scooping the sand used to create the house on the left. The house and the void complement each other as they have the same shape and size. If attention is given to the structure of a work such as *House at Nobby Beach*, and the process of its creation, even this simple work can become a source of many telling observations: the house is made with grains of sand and water; the water is likely to evaporate with time and the form will turn back into a shapeless collection of sand; to erect the house, the sand needs to be excavated and any water scooped out, which creates a void in the environment; and, once part of the house, the sand and water cannot constitute any other form; unlike the form they create, the grains of sand are eternal; and, because the water and sand can be reused continuously, the form could be thought of as a process; once created the forms slowly decay, so living could be understood as a maintenance of the form, and so on. One can already find in this early work all the elements which so captivated me during the last ten years: the issue of content, form and figure, scales, orientation, process and state, and the most difficult of all issues: the meaning of life.

FIG 2: RAYKOWSKI W, 2002, *ONE-METRE LONG CONSTRUCTION TIMBER*, PINE TIMBER, 180CM X 80CM X 6CM; NONEXTANT

The same concerns are also present in *One-metre Long Construction Timber*, which comprises the ash created by burning a one metre long construction timber. The importance of forms becomes apparent again: the object is not just any process, but a process which results in a lasting form. It is the form that matters;

the timber without its form is useless. This work presents the difference between life and death as one of having form, or not having it.

2 INVAGINATION

The most important outcome of the previous works was the realisation that forms are made of elements and space – without either there can be no form. This led to the investigation of how such forms, understood as spatial collections of elements, could interact. The first attempts involved the interaction between the form and a space external to it in the context of written text, as demonstrated in *Untitled (Invaginated figures series)*.

Fig 3: Raykowski W, 2002, *Untitled*, (Invaginated Figures Series), acrylic on canvas, 60cm x 120cm

Further works pushed the concept into non-textual realms. In *Narcissus,* the figure of a man and its reflection demonstrate two dramatically different ways of interaction between collection and space around it.

CONCEPTUAL UNDERSTRUCTURE OF HUMAN EXPERIENCE

FIG 4: RAYKOWSKI W, 2002, *NARCISSUS*, (INVAGINATED FIGURES SERIES), ACRYLIC ON CANVAS, 90CM X 60CM

The nature of the process is explained in Fig 5 below which shows how the same action directed at an unshaped collection in the middle results in two different outcomes. Think of a lump of clay which one moulds with fingers into the figure on the left which emphasises the positive aspect of the shape (a phallic representation) or the figure on the right (an invaginated representation) emphasising the negative aspect.

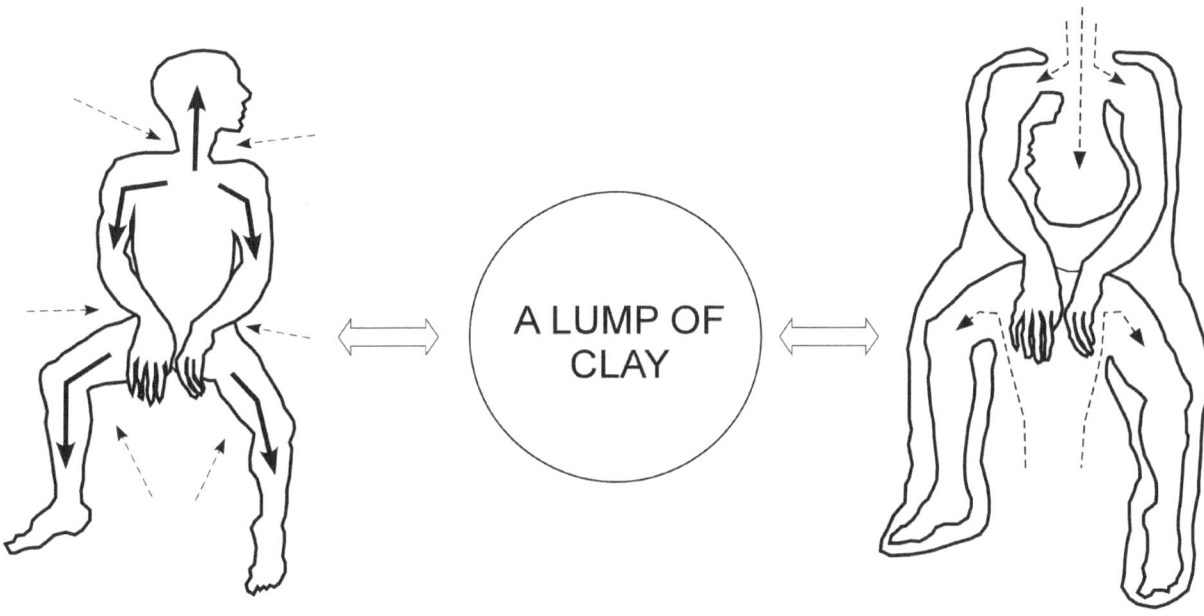

Fig 5: Phallic representation   Embryonic shape   Invaginated representation

The left figure draws attention to the scale from which the shape-forming space originated: the limbs of the figure protrude confidently into space (solid-line arrows) suggesting that their owner is in charge of his or her own life, actions and deeds: the head is the source of the individual's own thoughts; the hands work for the individual alone; the legs suggest a control of the individual's destiny and every step towards it; the ears, tongue and nose imply authority over what the individual hears, sees or says. The bigger and more defined the positive shape, the more significant its presence within the public space.

Conversely, the figure on the right draws attention to a scale smaller than the public space; to the internal space of the individual. Defined from without, the head of the individual is no longer the source of his thoughts but is filled, like a container, with another's ideas; the hands no longer work for the individual; the legs do not carry the body to a destiny that could be called the individual's own. The individual is no longer in charge of his or her own life - they are manipulated and used by an amorphous external environment that seems to have a life of its own. It is difficult to observe the impact of other people on the individual; the hidden effects can only be inferred from the behaviour of the person.

These figures not only convey the two distinct ways in which humans can observe themselves, but also draw attention to various scales and their interaction. The phallic shapes are not just visible but also powerful (large scale); and the invaginated shapes are not just hidden; they convey a sense of victimhood (small scale).

In *Red and White*, the invaginated form of an individual is entirely defined from outside, evoking a sense of helplessness and resignation in the orientation of the limbs and head.

CONCEPTUAL UNDERSTRUCTURE OF HUMAN EXPERIENCE

FIG 6: RAYKOWSKI W, 2002, *RED AND WHITE*, (INVAGINATED FIGURES SERIES), ACRYLIC ON CANVAS, 50CM X 50CM

The starting point for both the phallic and invaginated form is the sphere: an embryonic form where no part protrudes into space or is invaginated by it. The sphere is the most symmetrical of all forms. It seems to be independent from its background, as if isolated from it. Not surprisingly, the foetal position so often adopted when a person feels vulnerable resembles the sphere – it is a way of isolating the person from the immediate environment.

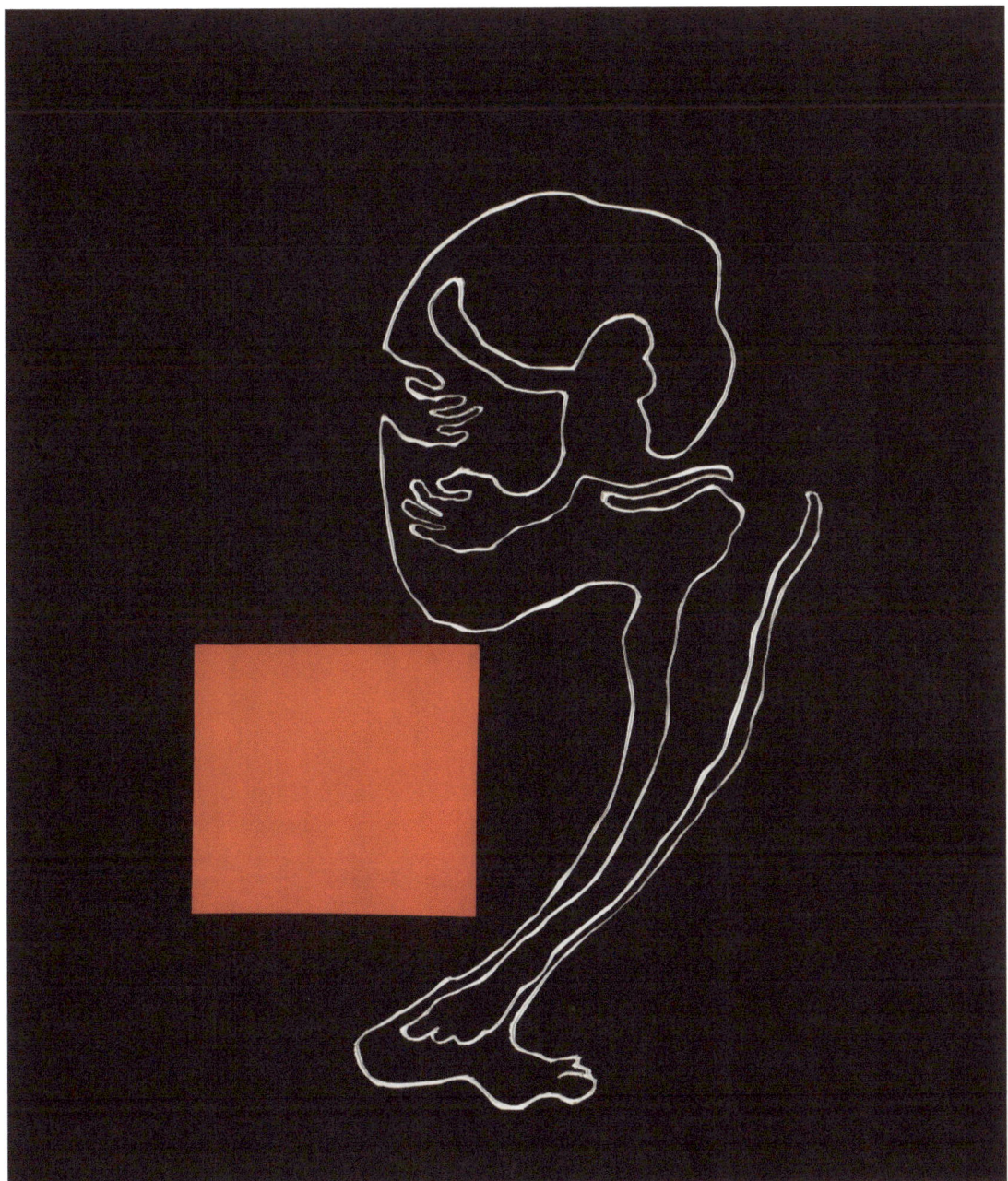

Fig 7: Raykowski W, 2002, *Red square*, (Invaginated Figures Series), acrylic on plaster board, 75cm x 88cm

Whether invaginated or phallic, all human figures have substance; they exist within a clearly defined space which makes the figure deformable. When the continuous boundary is ruptured, the space is no longer divided into the inside and the outside – what remains is the surface itself. A figure which has lost its substance is no longer an individual: *Standing up* illustrates a figure constructed from a single line, where all parts are defined entirely from without.

CONCEPTUAL UNDERSTRUCTURE OF HUMAN EXPERIENCE

Fig 8: Raykowski W, 2002, *Standing up*, (Invaginated Figures Series), acrylic on plaster board, 55cm x 120cm

It is no longer a figure of the form, only a shape of line, and as such has no life of its own. It cannot exist without an outside support; *Black rectangle and a thick line* illustrates this concept.

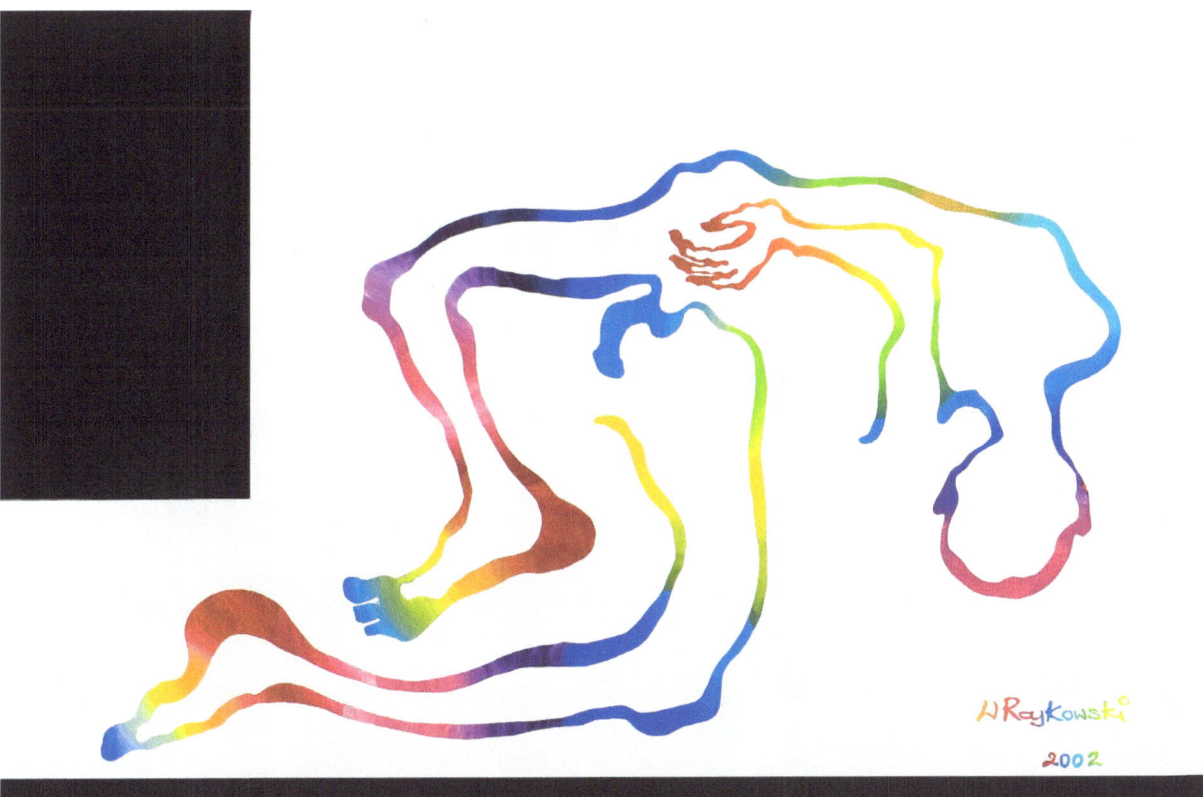

Fig 9: Raykowski W, 2002, *Black rectangle and a thick line*, (Invaginated Figures Series), acrylic on canvas, 90cm x 60cm

The content of the figure is no longer intrinsic; it comes from outside, and thus can be easily changed. The following chapter explores this concept and its attendant issues further.

3 CONTENT

Instead of any purchased products one might expect to find in it, the shopping bag in *A bagful of air* owes its characteristic shape to the action of a simple fan. The oscillation of the fan and rattling noise of the plastic adds to the sense of compulsion and futility associated with the shopping experience.

CONCEPTUAL UNDERSTRUCTURE OF HUMAN EXPERIENCE

Fig 10: Raykowski W, 2004, *A bagful of air*, (Invaginated Figures Series), plastic bag, fan, plastic crate, and label; 90cm x 60cm x 60cm

The work *Every time you enter the gallery, the air of your volume leaves the room* comprises a collection of seven plastic containers arranged into a shape which thrusts vertically into the space of the gallery. The total volume and height of the containers was calculated to represent the size and volume of the average Australian adult. The work was positioned just inside the entry to the art gallery. The label, which draws attention to a physical process most of us rarely think about, disrupts any expectation the viewer might have about the sculpture in the context of the gallery – promoting an awareness of the consequences of their own actions on entering the room.

Fig 11: Raykowski W, 2005, *Every time you enter the gallery the air of your volume leaves the room*, seven plastic boxes and label, 157cm x 35cm x 22cm

For the viewer, the space is suddenly no longer empty – it is a collection in its own right. The work highlights the hidden complexity of an apparently empty space.

4 DEFINING SPACE

Whether we are aware of it or not, all objects carry social connotations. To appreciate the role of space in defining relations between objects, one need to minimise such connotations; removing an object from within the space is probably the most effective way to achieve this. But this creates a problem: no spatial relations can be defined without the object. In 2004, I tried to overcome this conundrum by using labels instead of paintings in a large exhibition of works. The labels themselves were priced at nearly thirty thousand dollars (Fig 12).

Fig 12: Raykowski W, Show: *AU$29 300*, signage on the front of the gallery

In using tiny labels instead of large paintings, an acute awareness of space is born. To suggest the paintings, the labels were positioned precisely, not only in relation to each other, but also in relation to the floor and walls. The space between the labels, and to some extent the lighting, is the only thing that suggests the size and position of the works (Fig 13).

Fig 13: Raykowski W, Show: *AU$29 300*, overall view

Fig 14: Raykowski W, *Painting that doesn't exist no 4* (Show: AU$29 300), medium undetermined, size unknown

The non-existing paintings were priced for quick sale, the cheapest as low as $50 (well below the cost of production) to make them affordable to the general public. Even so, only one patron was prepared to buy an artwork: a doctor at the local hospital (let us call him Professor P) who was an avid collector of *art brut* (outsider art). Selling non-existing objects is not a straightforward business; one has to come up with a tangible way of conducting the trade. I made all necessary arrangements at the gallery to document the act of selling the artwork, but the buyer never showed up to complete the transaction. I later learned from other sources that under pressure from his peers at the hospital, Professor P changed his mind about acquiring the artwork. I never met him again. This incident made me acutely aware how much the notion of *value* depends on society: the sense of quality might come from within the individual, but its acceptance is determined entirely by the social group and its culture.

CONCEPTUAL UNDERSTRUCTURE OF HUMAN EXPERIENCE

5 INVESTIGATING HEIGHT AND WIDTH

Suggesting the nonexistent paintings by positioning labels on the wall can indeed be a very tangible activity – the distance between the labels has to be defined precisely in terms of width and height. The work *Distance from the ceiling to the plinth* in Fig 15 below looks into the latter notion.

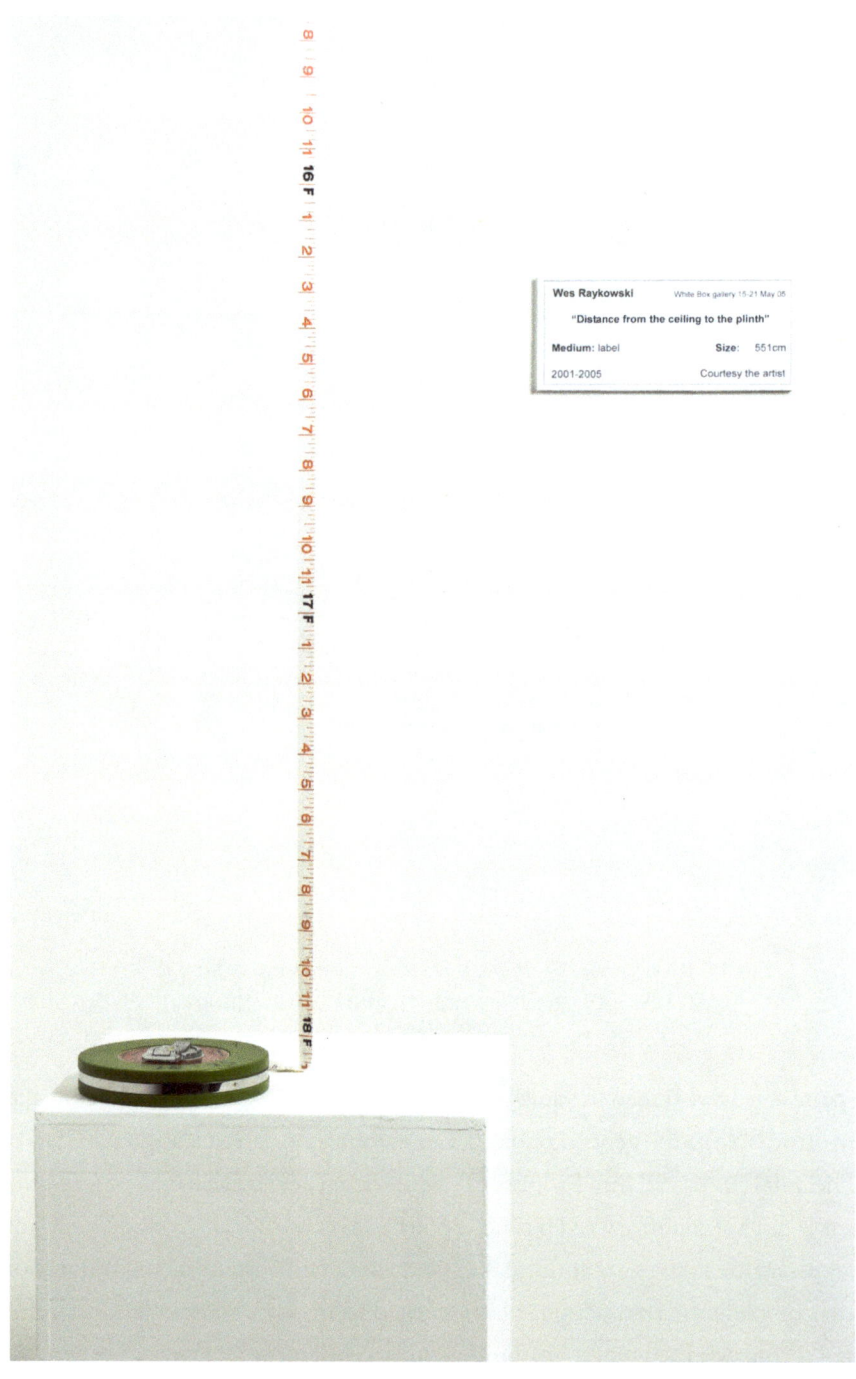

FIG 15: RAYKOWSKI W, 2002, *DISTANCE FROM THE CEILING TO THE PLINTH*, A LABEL, 20 METER-LONG TAPE ON PLINTH; SIZE 551CM

The title *"Distance from the ceiling to the plinth"* suggests that the subject of the work is not the measuring tape itself but the distance of 551 cm which the tape conveys. The artwork is significant in that it draws attention not only to the fact that the concept of distance cannot be defined without an object such as a measuring tape, but also in that the object used to measure is itself structured in terms of the space it is supposed to define: the tape is created by dividing its length into smaller-scale spaces such as meters (each marked with two object-like lines), which are defined with even smaller spaces such as centimetres (each marked with two object-like lines), potentially without end.

Before my involvement in fine art I thought, just like most people, that height and width are just different names for distance because they can both be described in terms of the same measuring procedure. This impression was undermined by the painting *"a little more to the left"*. The work makes the viewer aware of comparison as the procedure which in many ways is opposite to the process of measuring depicted in Fig 15.

Fig 16: Raykowski W, 2005, *"a little moret to the left"*, acrylic on canvas, string, a nail, nail holes and label; 140cm x 84cm

The subject of this work is not the image on the canvas, which is empty, but fifteen or so nail-holes in the wall which are the only evidence of my attempt to position the painting at the desired location. If the experience of placing an object at the right location is analysed, however, one becomes aware that comparison cannot be carried out without the individual who makes the comparative judgement - what is compared exists in the space outside of the person, but the comparison itself takes place always inside that individual. The work also shows that comparison can involve progressive refinement – the holes are placed closer to each other until the right location is found. Once again we are confronted by the presence of scales in which two processes occur: the measurement takes place at the scale of the gallery where the individuals and objects are related by space; and comparison occurs at the smaller scale of the individual's own interior. What is more, the procedure of measuring relies on the process of sensory comparison without which juxtaposing the object used to measure against the object to be measured would remain uncompleted.

6 LEVELS, HEIGHT AND COMPARISON

The next two works demonstrate that comparison defines the relation between objects in terms of the space between them; and the process of measuring is nothing more than the comparison of two objects, one of which happens to be already predefined in terms of space. By creating these and other similar works I also realised that *height* is typically associated with comparison and *width* with concatenation, which is a fancy word which may be used in place of 'measurement'. This observation suggests that height is usually assessed in terms of the judgement made by an interpreter (something is higher than something else without specifying how much) and width is specified in terms of the unit used in the course of measuring (e.g. distance is a multiple of the distance unit and weight is a multiple of the mass unit).

Consider the painting titled *"The artist's height as marked on 7 July 2001"* below. The work depicts the common practice of recording the process of growing up by marking the height of a person on a vertical surface such as a wall or a door jamb. If the mark is made on the canvas instead of the wall, however, the record is likely to become meaningless when the canvas is shifted away from its original setting. This work draws attention to the fact that the concept of height is independent of the measurement as it relies solely on the process of comparison which depends on a stable relationship between the floor and wall. Because canvases can be easily shifted, they do not provide the best surface for recording someone's height.

Fig 17: Raykowski W, 2001, *Artist's height as marked on 7 July 2001*, acrylic on canvas and label, 50cm x 50cm

But here is the twist: the canvas is as good a surface as the wall for marking the progress of a person's development as long as the painting is kept in the same position during subsequent recordings. If, after a few years, the canvas is shifted to another location, it only will lose the record of the actual height not the information about its change. Drawing two lines on the canvas instead of one, therefore, would make the artwork less challenging and, more importantly, less humorous. Such a painting will be no different to a measuring tape without an origin.

Measuring the height of a person is relatively easy. But try this in the case of a mountain. Horizontal distances, on the other hand, are always easier to measure irrespective of their length - if they are really large, such distances can be measured with steps. An example of the work *Ten feet long tape,* which shows an image of a ten-feet-long sticky tape on which the artist carefully stepped.

Fig 18: Raykowski W, 2004, *Ten feet long tape (footprints of the artist walking over the sticky tape)*, adhesive tape on the wall, 267cm x 4.8cm, nonextant

The human feet can be used to measure flat surfaces by placing one foot in front of the other. But even easier than using feet is making the measurement with steps as it is impossible to separate two subsequent steps simply because the leg that ends the last step is always the same leg that starts the new one. One can say that steps, unlike feet, are always perfectly concatenated. Therefore, as long as we can walk across or along it, any object can be measured. It may not be the shortest distance, but it is always meaningful to the person who measures it. Unlike in the case of distance, there can be only one height. This is because, to measure an object's height one needs to have access to its base, which is not always possible. In such situations, the process of indirect comparison is used: the object of unknown height (e.g. a mountain) is compared with another object, the height of which is certain, to find out how many times one is larger than the other. The comparison is made in terms of the viewing angles as illustrated in Fig 19 below.

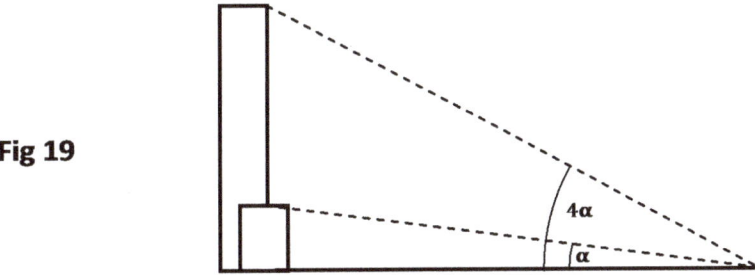

Fig 19

For this to be possible a person is needed who can conduct and interpret the process of comparison by sighting both objects. Once again, the process suggests the presence of two adjacent scales: the two objects might exist outside (larger scale), but the process of angular comparison between them takes place firmly inside the interpreter (smaller scale). The sensory experience of such a process was vividly captured in the work "*a little more to the left*" in Fig 16.

7   EMERGENCE OF LARGER SCALE

The notion of scales can be confusing – it is usually understood in terms of the size of an object rather than the structure of a collection which results in a particular size. When making an artwork, it is difficult to avoid the issue of scale in the first sense: the reproductions of figures are typically scaled down and sometimes enlarged or left unchanged. The artworks dealing intentionally with the second sense of scales are less common in my oeuvre. I can think of three works that illustrate this point; two of which are discussed in this section.

The oval paintings below exist at two scales: as a collection of four distinct paintings; and as a single work united by the image of the rhombus with oval corners. Space plays an important role in this work: separate the smaller paintings further apart and the emergent figure will become much larger, and vice versa. Together with the elements the space creates the larger-scale form.

CONCEPTUAL UNDERSTRUCTURE OF HUMAN EXPERIENCE

Fig 20a: Raykowski W, 2004, *Oval section 1*, acrylic on canvas, 20cm x 25 cm

Fig 20b: Raykowski W, 2004, *Oval section 2*, acrylic on canvas, 20cm x 25 cm

Fig 20c: Raykowski W, 2004, *Oval section 3*, acrylic on canvas, 20cm x 25 cm

Fig 20d: Raykowski W, 2004, *Oval section 4*, acrylic on canvas, 20cm x 25 cm

The distance between elements is therefore meaningful in two ways – it defines the relation between the elements (smaller scale) and the overall size of the work at the larger scale.

The work *Society is what becomes of individuals when they get together* deals with the social aspect of scales – the notion of individuality and social group. The artwork is an interpretation of the Spencer Tunic photograph titled *Mexico City 1 (Zócalo, MUCA/UNAM Campus) 2007*. Tunic is known for using human beings "stripped down" of their own individuality (by requiring them to remove clothing and adopt the same position) to create and photograph large-scale installations made of anonymous elements. In the case of this work, my intervention is confined to the title which draws attention to the individuals' willingness to forgo their own personality in order to participate in the activity of the group.

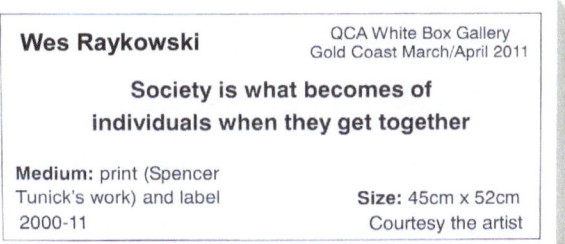

FIG 21: RAYKOWSKI W, *SOCIETY IS WHAT BECOMES OF INDIVIDUALS WHEN THEY GET TOGETHER*, PRINT (SPENCER TUNICK'S WORK) AND LABEL, 45CM X 52CM

Spencer Tunick's installations are of interest to me as an example of the human craving for order (in the sense of predictability) so characteristic of the highly controlled environments we find at workplace, universities, army and churches. For the worker, student, soldier or a believer, a day-to-day life might be very predictable, but the effect the group has on the physical and social environments are usually not. Stripping can therefore be more than removing clothing – it means giving up the ability to judge and with it control of one's own destiny.

8 INDIVISIBILITY OF UNITS

In addition to space, the elements are another essential ingredient without which no form can exist. The subject of the next four works is the issue of units and their structure and relation to larger scale form. In the context of my art practice, I consider the elements to be synonymous with units, but see them as referring to different scales: units bring up the small scale and elements the large scale. Perhaps the best way of examining units involves dividing them into parts. In work *Two sides of the coin* I cut a two-dollar coin into halves.

FIG 22: RAYKOWSKI W, 2004, *TWO SIDES OF THE COIN (SEPARATING HEADS FROM TAILS)*, ANXIETY SERIES; AU$2 COIN SPLIT IN HALF, 2 X (2CM X 2CM)

In another work I erased one side of one-dollar note (Fig 23).

Fig 23: Raykowski W, 2004, *Erased dollar note (Erased face of US$ bill no B90077927 A)*, Anxiety series, US$1 note on board, 5,7cm x 6.6cm

In yet another work I glued two one-dollar notes together (Fig 24).

Fig 24: Raykowski W, 2004, *Two dollars (Two One Dollar notes fused together face to back)*; Anxiety series, note no E38927281 F and B91870067 D; 15.7cm x 6.6cm

And there is also a work in which I cut the measure tape along its length (Fig 25).

Fig 25: Raykowski W, 2004, *A half of a metre (One metre long measure tape cut in the middle along its entire length)*; steel measure tape, 1 metre

Each time I pondered on the value of the manipulated unit. The forthcoming answers were not clear. One half of the tape could still be used for measuring while the other half cannot. On the other hand, when glued together, notes become worthless; and whichever way the coin is cut, it loses its value entirely – a half of the two-dollar coin is not one dollar and a fifth of one-dollar coin is not twenty cents. This would not be the case, however, if the coins were made of gold rather than copper alloy. A half of the gold coin is likely to retain a half of its value. The answer is then not that straight forward: in some circumstances units could be cut and in others they cannot; but why? Unlike the alloy coins, the gold coin is a collection of atoms of gold which constitute the smallest indivisible unit of the metal from which the coin is made. As the unit of tender, the gold coin is divisible, but not its atomic units.

Divisibility is therefore what defines all units and elements: a unit is the smallest indivisible element that other things are made of. One cannot cut the alloy coins in half and hope for parts to have the proportional value, the same way one cannot divide the atom of gold or cut the cell or a human being in half and hope that they will remain fully functioning units. Atoms, cells and human beings are spatial forms not just aggregations of elements. As forms, all such units are indivisible. Playing with coins, notes and a measure tape at art school can be more than just poking fun at society and its values.

9 CONTRASTS

The notion of contrasts and its role in cognition is one of the most important issues discussed in the thesis. I explored contrasts in a number of works, three of which are included below. In the work called "*I hate painting*" (Fig 26), I recorded the act of painting a written statement on a white wall using white paint. Once the masking tape was removed the sign became invisible due to absence of contrast.

CONCEPTUAL UNDERSTRUCTURE OF HUMAN EXPERIENCE

Fig 26: Raykowski W, 2005, *Untitled (I hate painting)*; a half-hour-long performance; acrylic on acrylic painted wall, masking tape, video player on plinth

In another work titled *13 14 15 22 28 30 32 38 40 – 26 28 37 41 43 44 45 – 3 5 7 8 14 23 31 39 – 2 4 6 8 9 18 27 33 36 41 43 45 – 1 2 11 20 29 36 38 40 41 43 – 34 36 38 40 – 1 3 5 7 9 17 26 35 38 39 – 1 3 6 33 35 37 39 – 33 35 37 39 – 6 13 20 27 33 34* I created contrast indirectly by listing the numbers which refer to the grid provided by the lotto coupon. Please note that the games 3 and 4 are not used in this depiction.

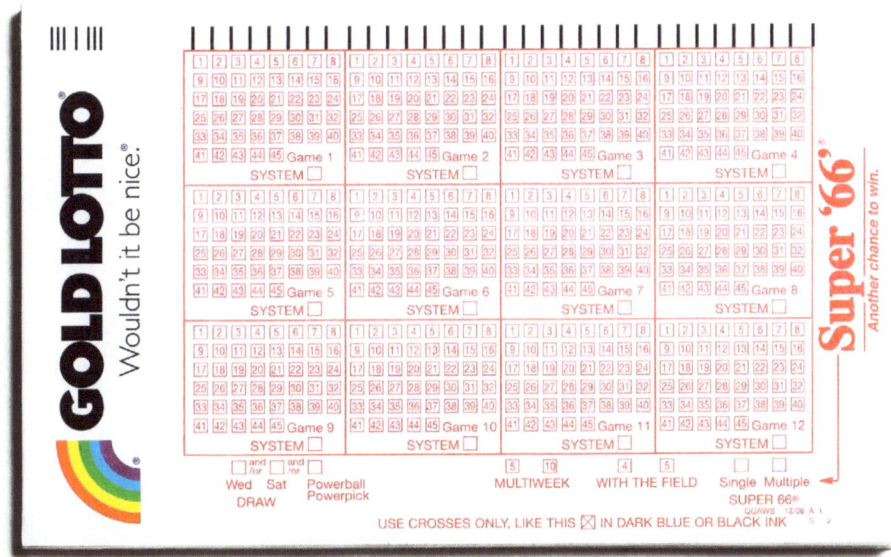

Fig 27: Raykowski W, 2007, "*13 14 15 22 28 30 32 38 40 – 26 28 37 41 43 44 45 – 3 5 7 8 14 23 31 39 – 2 4 6 8 9 18 27 33 36 41 43 45 – 1 2 11 20 29 36 38 40 41 43 – 34 36 38 40 – 1 3 5 7 9 17 26 35 38 39 – 1 3 6 33 35 37 39 – 33 35 37 39 – 6 13 20 27 33 34*", lotto coupon and label, 18,5 cm x 26 cm

The sense of contrasts as suggested in this thesis is made more palpable in the more recent work called *$40*, in which nearly four thousand ¢10 coins are arranged on the floor into a rectangle (Fig 28). Initially all the coins were placed on the floor with their heads up. The title is derived from the image created by turning some of the coins around. Because the two sides of the coin do not differ significantly when looked from a distance, the image is practically imperceptible. As in the case of *I hate painting*, the image is there, but it cannot be seen due to insufficient contrast.

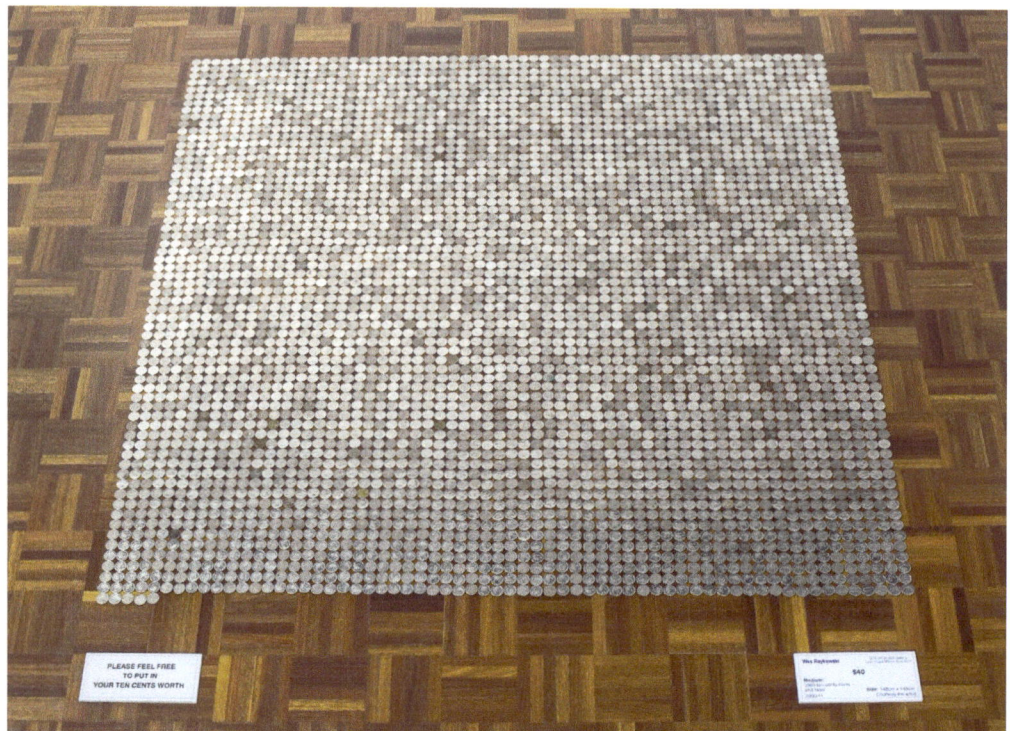

Fig 28: Raykowski W, 2011; *$40*; 3969 ten-cent coins and label; 148 cm x 148 cm

This works alludes to the process of creating contrasts in the cortical maps, some elements of which are activated while others are not. From the outside, the activated cells do not differ from inactive cells. The image in such a situation can be appreciated from within of the collection only.

10 LINES AND THEIR ORIENTATION

One of the returning themes in my artworks is the issue of line orientation and body position: why is it that diagonal lines seem to convey movement; vertical lines: happiness, intensity and energy; and horizontal - the absence of energy, resignation and death. Why is depression often depicted with a mixture of diagonal, vertical and horizontal elements (as in the case of the foetal position), but joy is represented by diagonal and vertical lines only (as in jumping for joy)? The usual answer is that artists paint or draw what they see and the lines just happened to be this way. According to this view, there is nothing special about the orientation of the lines, so there is nothing to question or interrogate. I felt otherwise and investigated the matter in depth, with experiments such as *Calm Sea*.

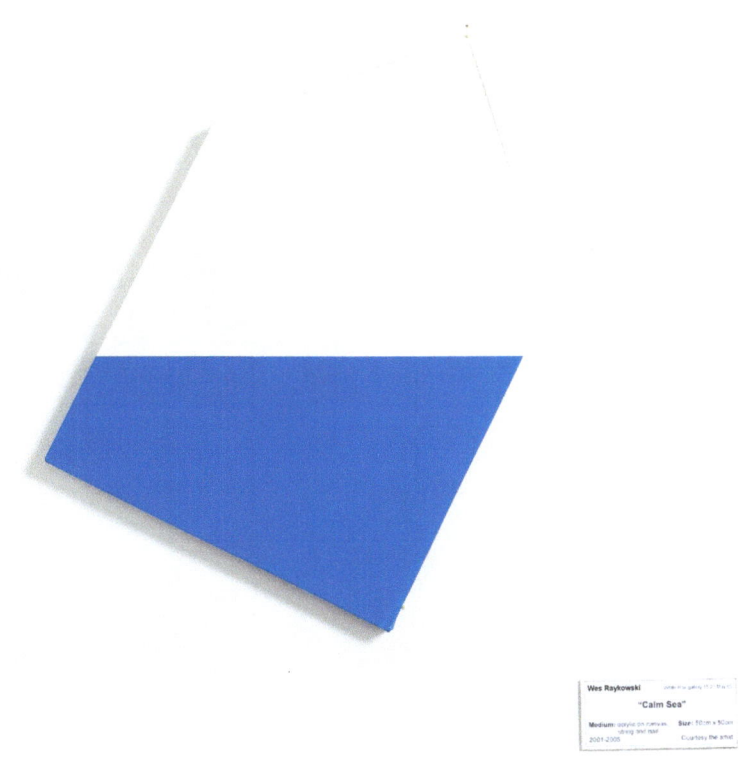

FIG 29: RAYKOWSKI, 2005; *CALM SEA*; ACRYLIC ON CANVAS, STRING AND NAIL; 65 CM X 70 CM

At first sight, the work might seem simplistic, if not primitive – the painting is tipped over to suggest the level of the titular sea. The painting is rotated, but the surface of the sea is not (Fig 30): the sea seems to exist within the frame of the viewer (*axes I-E*) instead of the frame of the painting (*axes i-e*). This violates the viewers' expectations – if the subject matter were part of the painting, its image should be rotated together with the painting. Some viewers might even try to correct the misalignment by rotating their heads, but such an act does not help as the sloping water contradicts the title which insists that the sea is calm. The issue rarely arises as the two frames are usually aligned.

By misaligning the frames, the work makes the viewers aware of their own orientation and helps them to observe the internal structure of their poses.

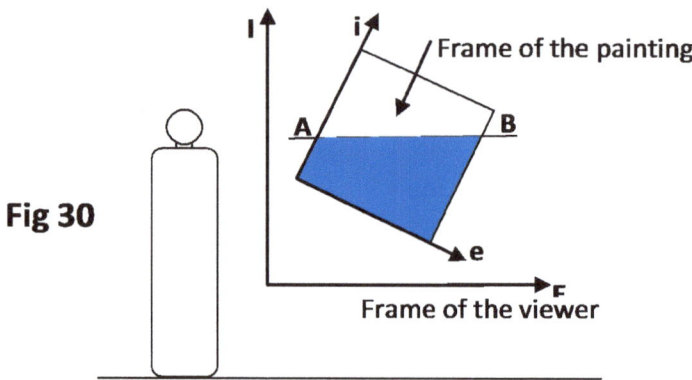

The notion of value associated with the vertical orientation is investigated in *Berlin Street Scene with Red Streetwalker*; *Yellow Square(M12)*, *Still life with Mortar, Jug and Copper Saucepan*; a series of small paintings described in greater detail in Appendix B and displayed in close up in Fig 31.

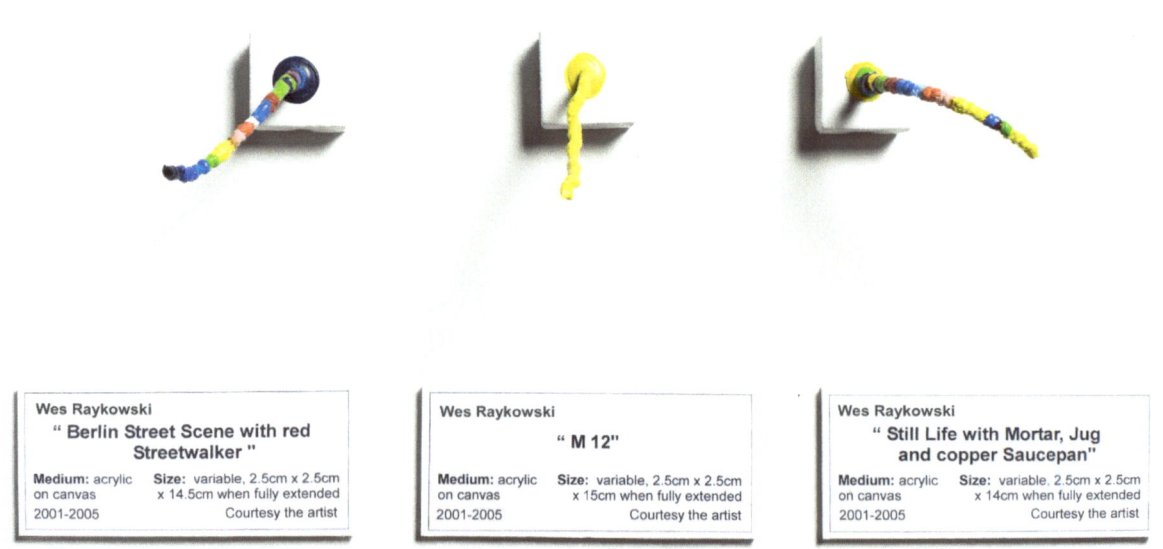

FIG 31: (FROM LEFT TO RIGHT) BERLIN STREET SCENE WITH RED STREETWALKER; M12, STILL LIFE WITH MORTAR, JUG AND COPPER SAUCEPAN; 2001-2005, ACRYLIC ON CANVAS; 2.5 CM (WIDTH) X 2.5 CM (HEIGHT) X 15 CM (DEPTH)

The works are highly significant for the following reasons:

First, despite of their seemingly tiny size, creating such works takes a lot of time – the beads of paint cannot be added before earlier blobs are relatively dry. Because of the long construction period, there are plenty of opportunities to experience the process and analyse the experience.

Second, the process is highly repetitive, which helps the observation and aids the analysis.

Third, the built up of the paint is very sensitive to deviation from the vertical orientation – if not careful the protrusion will lean or even collapse. The problem gets worse with the mounting height of the protrusion which makes one acutely aware of the intensive aspect of the work.

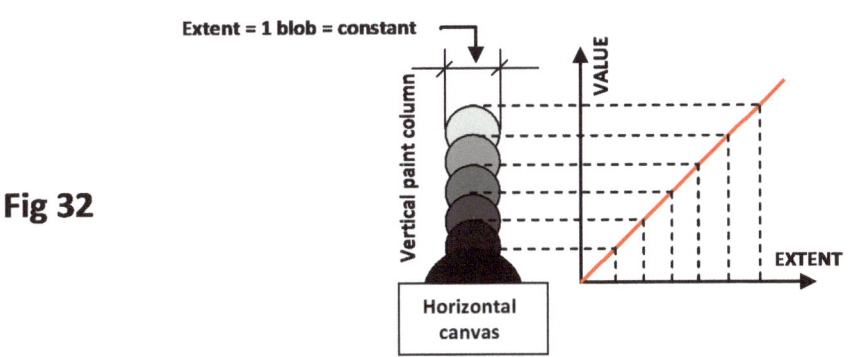

Fig 32

Fourth, there is a sense of increasing value with each blob of paint added to the painting.

Fifth, even though they represent different aspects – intensity involves the sense of effort and value the promise of social recognition – the two notions are connected through the work.

Sixth, if all the blobs of paint are of the same size, the extent of the sculpture becomes fixed during the course of its construction. Fixing the extent draws attention to the process of comparison.

Seventh, if more than one painting is created, one starts to appreciate the notion of extent as that of repetition of the same unit.

Eight, when the protrusions are hung on the wall, the sense of intensity and value associated with the vertical orientation turns into the notion of extent associated with the horizontal orientation, height is converted into the length, and the process of comparison is replaced with the process of repetition.

Ninth, with time the protrusions, placed in the horizontal position, will sag, which is perceived as a loss of potency associated with the steady increase of the intensive component in the direction of sagging (refer to Appendix B for a demonstration).

The interaction between the intensive vertical and extensive horizontal component of diagonal lines was studied in works which utilise the string on which the painting usually hangs. When brought out from behind the canvas and stretched between nails, the string can be used effectively to explore the changing relationship between line sections having different orientation, and to analyse links between the social meaning of the figures and the orientation of the lines that makes them. One example of such a work is *Four arrows*.

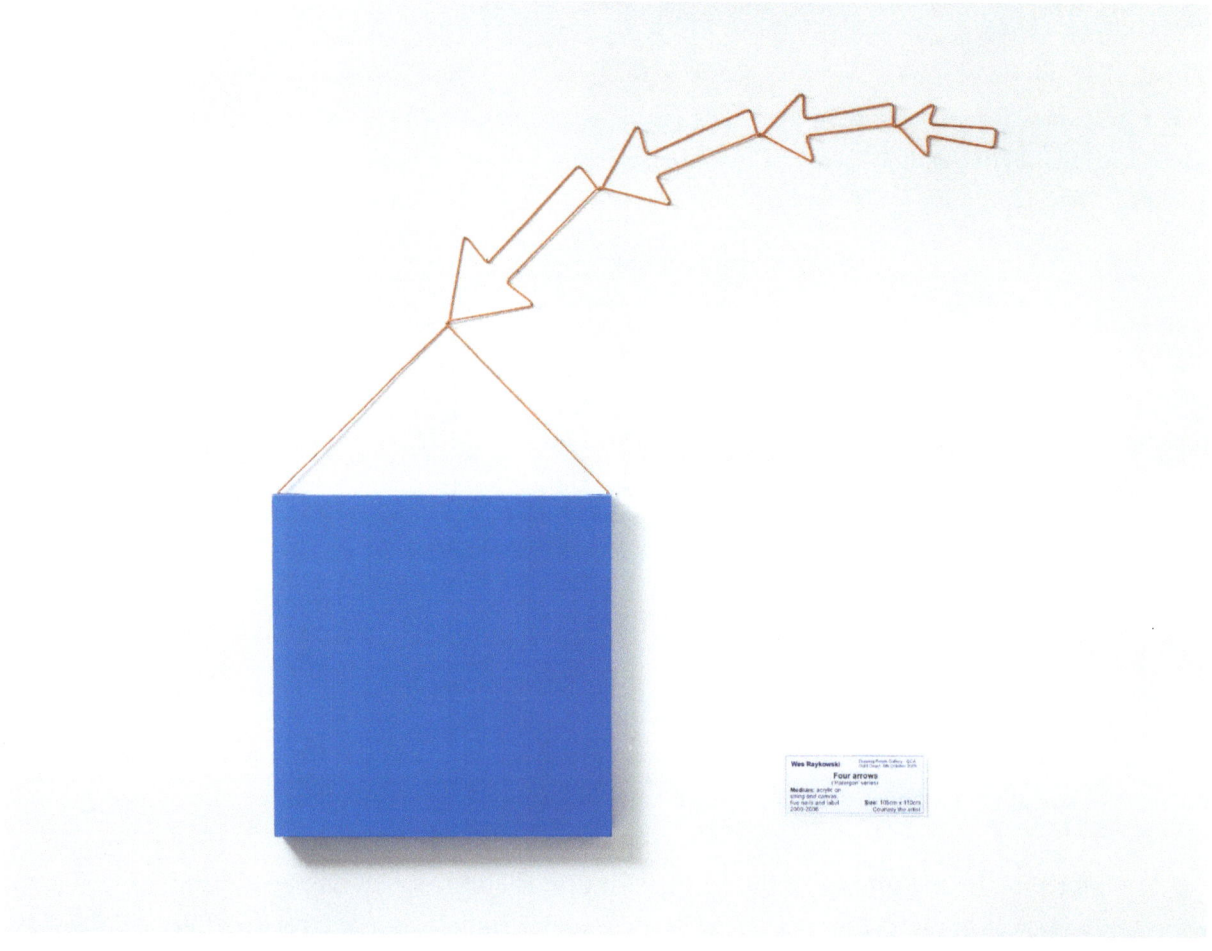

Fig 33: Raykowski W; 2005; *Four Arrows;* acrylic on canvas and string, nails and label; 105 cm x 110 cm

When considered within the frame of the viewer (Fig 34), any change of the orientation of each arrow could be interpreted as the change of its vertical (intensive) and horizontal (extensive) components. The far-right arrow suggests a modest upwards movement because of the small but positive direction of its intensive component; and the remaining arrows suggest an increasingly downwards motion because their intensive components are negative and fast increasing.

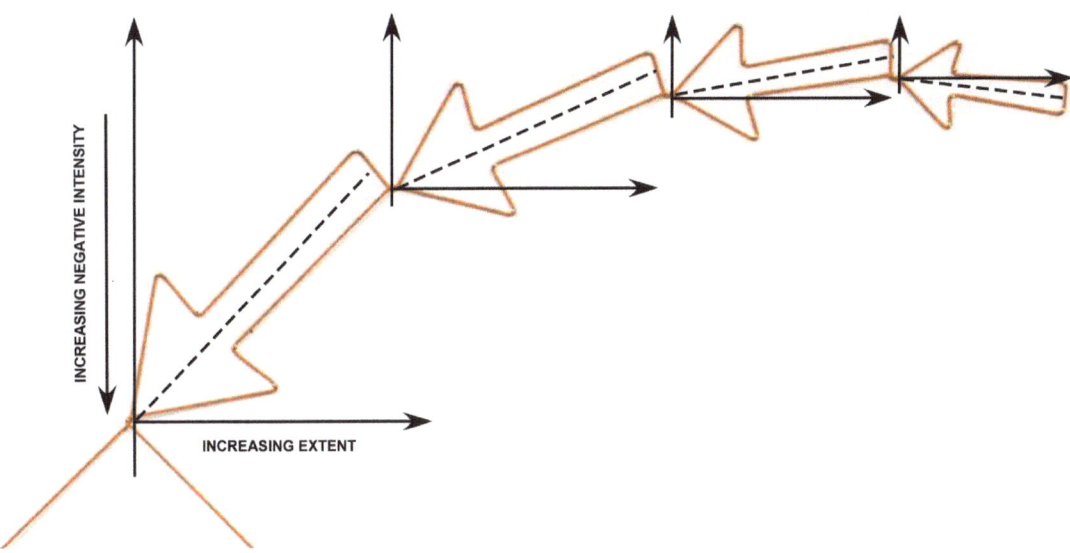

Fig 34: Detail of (Raykowski W; 2005; *Four Arrows;* acrylic on canvas, nails and string; 105 cm x 110 cm)

This line orientation is also significant in the case of the three-dimensional representations of solids which is an issue investigated in *Text*.

CONCEPTUAL UNDERSTRUCTURE OF HUMAN EXPERIENCE

Fig 35: Raykowski W; 2011; *Text;* acrylic on canvas and string, nails and label; 60 cm x 120 cm

In this and many other similar works, the canvas is used not for its pictorial space, but to tension the string. To remove any possibility of interpreting the canvas as an image, the string was painted with the same paint as the canvas.

The three-dimensional aspect of line orientation was also investigated in *Chair*, (Fig 36) which makes a playful use of the canvas as a seat of the chair.

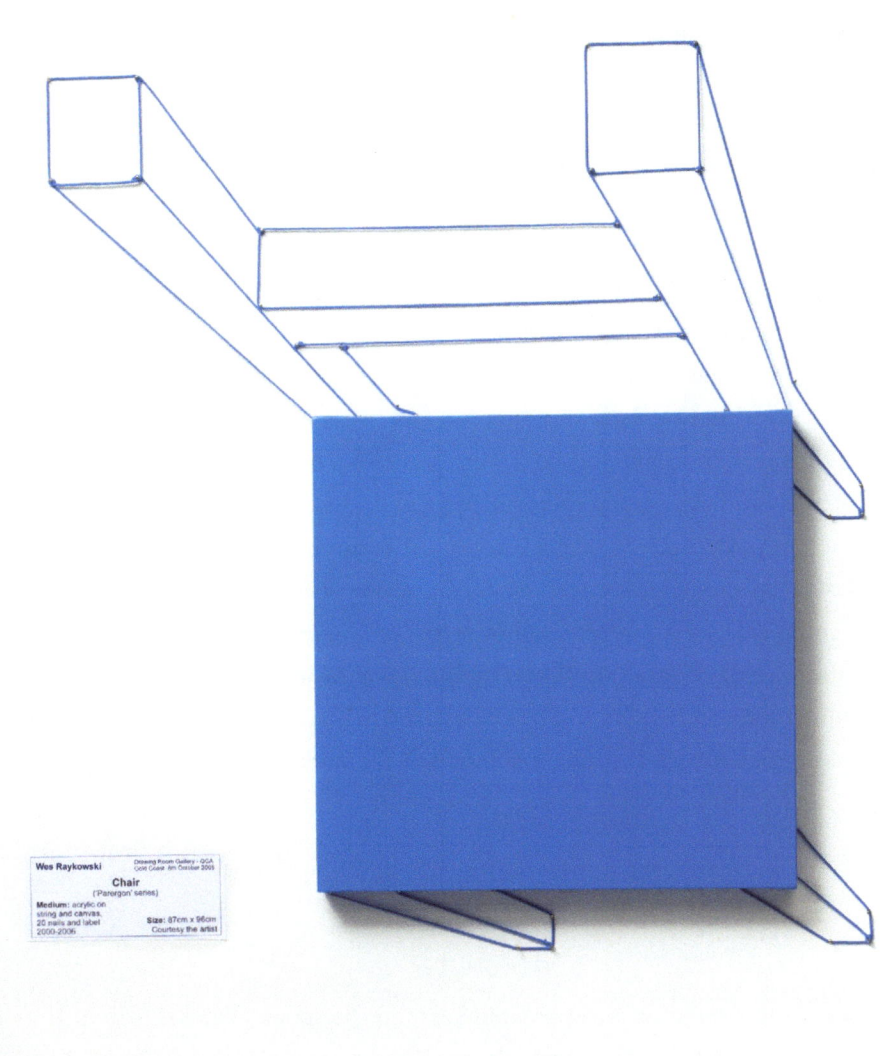

Fig 36: Rąykowski W; 2005; *Chair;* acrylic on canvas and string, 20 nails and label; 87 cm x 96 cm

My study of images and sculptures has confirmed my earlier observations that the concept of orientation is highly significant for the human experience and all its expressions. Elongated elements do not just happen to be oriented in a particular way – their orientation reflects the structure of the human body and its physical environment, the interaction with which results in a cognitive schema that combines the experience of intensity (or value) with the experience of its extent. Such elements are capable of representing movement or its absence precisely because they have components that indicate a degree of intensity (or its absence) and its extent. With these components combined in distinctive ways, humans can express their resignation or hope, woe and joy, anxiety and relaxation, or even the feeling of death or living. This capacity is common to all people because they all share common anatomy, physiology and the physical environment.

The same cognitive structure could be found in all expressions irrespective of their form. The notion of intensity and extent, hence orientation and its change, is used to organise and interpret colours and their shades, the general composition of the image, and the relation between its parts in the case of images, sculptures and installations. The orientation of elements helps us interpret not only drawings and paintings, but also diagrams, graphs, plans and charts. The orientation is also significant for language where the intensive aspect of phrases and sentences is expressed with adjectives and the extent with nouns complemented by determiners; and where verbs, which describe the intensity and extent of the change, are complemented by adverbs. The orientation could also be identified in human experience of music and the understanding of its records – where pitch of the tone is interpreted as intensity and duration as its extent; and where higher tones of the scale in music notation are placed at the top of the stave and lower ones at its bottom, and duration is expressed by specialised symbols standing for repetition, and so on. Physics too exhibits the same structure – except for a few dimensionless constants, most physical concepts are always expressed as some combination of intensity and extent, and their properties are described as intensive or extensive depending on their reliance on the system's size. And contrary to the common perception that mathematics is abstract and disembodied, such concepts as products, ratios, Cartesian coordinates, dimensions, fractals, graph and group theories, or even sets are all interpreted in terms of the intensive and extensive aspects of human experience. Even the everyday objects are always built to take the intensive vertical and extensive horizontal qualities into account. None of this should come as a surprise if one takes into account that the source of all concepts, the human experience, is organised in terms of the intensity of sensation and its extent.

11 CAN ART MAKING BE A RESEARCH

It is difficult to answer this question without first defining the scope of the research. If its topic is cognition, which is the area of interest to me, then the answer has to be a qualified yes. Art making is by no means a foolproof or easy research methodology, however. As any social activity, art too involves institutional pressures to perform a range of activities defining this field. In the case of fine art this means producing ever more artworks and exhibitions, which leaves little time for reflection involving the experience of creating an art object. Without time to reflect it is difficult to develop the skills of self-observation and without such skills it is impossible to make sense of the experience. In the situation when the process of creation is ignored, the objects have to be interpreted in the social context which is learned rather than directly experienced.

Making an artwork is not the same as creating an object; this is because the artwork is more than a shaped piece of material. It is a way of *thinking with one's own body*, but only if the artist will allow himself to observe the experience of the creative act. Art making can be a slow and often arduous process, especially if it is not automated by skills. And slow processes give plenty of time to observe. An automation of the behaviour, which comes with skills, makes the process of creation unavailable to direct introspection and subsequent reflection. Acquiring skills, therefore, robs the maker of the ability to experience such processes. Skills are not essential for creation. They are necessary, however, for manufacturing exact copies to satisfy the need for products. But because some artworks might be in demand it does not mean they should be recreated. This is because artworks are more than just objects and art making more than the process of manufacturing them.

Making an artwork (or any handmade artefact for that matter) combines the human body with its mind, and the physical aspect of human existence with its intellectual side. Contemporary art practice is naturally suited to such an enterprise as it provides room for all forms of human expression. Creating images is one way which is not without its problems – a picture might be worth a thousand words, but not a word more. What is more, the images just as all objects are particular in the way they can express experience. Unlike words which can stand for anything they do not resemble, sculptures and images are very specific. As a result, they can only represent their own copies. When combined with words, however, the experience conveyed by images and sculptures becomes more complete hence easier to understand – the words provide the context for the specific objects which in turn compensate for the generality of symbols.

Art making and scientific research differ in many respects. Their methods and objectives are in some ways contradictory: if the purpose of art is making intangible ideas real, then changing reality into a system of ideas is what science does. Art materialises abstract thoughts, feelings and perceptions, and science dematerialises what is by its nature material. Science, therefore, imposes constraints on human thought which art tries to shake off. Yet, just like science, all art is constrained by the physicality of the material world within which they both exist; and science is restricted by the human ability to experience such a world and make sense of it. The two fields are naturally combined into a single system by the human body which makes possible for humans to investigate art with science and study science with art.

There are no doubt other ways of viewing art as a research methodology which I have not explored, experienced or cared to mention in this thesis. What I described so far is only an interpretation particular to my experience of art. In the context of my investigation, the scientific aspect of my artistic research is clearly a by-product of art practice: I start my investigation with an artwork made as a result of an emotion; observe the process of its creation and, some time later, analyse this experience in a search for patterns which allow me to formulate a hypothesis in the form of a conceptual schema. This is where the process significantly diverges from the scientific method. Instead of designing experimental studies to test such a schema, I investigate expressions made by other individuals (e.g. the structure of a hypothesis itself and its place in theory) in search for the structures I identified in my own experience. If such schemas occur repeatedly and regularly in a large number of related expressions, I conclude that they are part of the conceptual repertoire of the individuals who made those expressions. It is not a proof, therefore, but yet another observation suggesting that the schema is likely to be statistically significant. The most significant part of this process is the first step in which one becomes aware of the conceptual regularities in one's own sensory experience. For this to happen, however, one has to produce an artwork which is meaningful to the artist rather than audience.

In this light the interpretation of my oeuvre outlined in the earlier sections of this appendix has to be qualified: the comments represent a post hoc rationalisation of my artwork and the experience of its making. At no point of my undergraduate and postgraduate training was cognition a subject of my work. Instead, such issues as emotions, self-expression and later the critique of fine art and society in general dominated my works. There can be no doubt that my interest in psychology and the circumstances of my education clearly contributed to my subsequent interest in cognition. ■

APPENDIX B

Extract from the catalogue

***Collab*eratum: *UQCA* @metro**[2]

Fig 1: Wes Raykowski, (from left to right) *Still life with Mortar, Jug and Copper Saucepan*; *Yellow Square*; *Berlin Street Scene with Red Streetwalker*; 2001-2005, acrylic on canvas, printed instruction and labels; 2.5 cm (width) x 2.5 cm (height) x 15 cm (depth)[3]

---

[2] The exhibition took place in 2005 at Metro Arts in Brisbane, Australia. It was organised by The University of Queensland, Metro Arts, and Griffith University Queensland College of Art.

[3] The photograph shown in the appendix was taken by Wes Raykowski at the exhibition opening night. The catalogue describes the right-most item of three paintings in the photograph.

Wes Raykowski is one artist who takes his art and practice deadly serious. Whether it is his incessant search for the "right" phrasing of his artworks' titles or his explicit guidelines for the installation and care of his artworks, Raykowski's fervour is both passionate and meticulous. Although his method and style of painting (if Raykowski would allow one to describe it as such) has evolved significantly since he first began his studies at QCA, his cynicism and pessimistic concerns for society have not. With resolute attitude, Raykowski explores ingrained social mores and openly questions their inner workings and their destructive consequences for individuality. Raykowski's method and style for creating his "protrusions" removes any possibility of expressionistic and painterly gestures and substitutes them with strongly worded and emotionally loaded statements referencing the history of painting. His fascination with language is obvious and his titles become an integral component of the artwork. The titles are deliberately arranged to interact with the images and to complement them; as he says:

> *I use rather evocative titles, which contradict and support the work while at the same time make (concealed) references to art history. Can you guess which artists I had in mind? Almost each word, its position within the title and relation to other parts of the title, etc., has some meaning – I play with labels the way other artists play with paints.*

In accepting his challenge, one discovers [that one of] Raykowski's artwork[s] borrows his title from German Expressionist painter Ernst Kirchner's *Berlin Street Scene with red Streetwalker*, 1914. Comparing Raykowski's "multicolour protrusions" to the painting of Kirchner, one recognises his indebtedness to Kirchner's brilliant colour scheme and his highly emotional and subjective response to modern, urban stimuli and the inner emotional truth of objects, people and experience. Kirchner's original painting depicts a prostitute dressed in red, surrounded by anonymous beings, with mask-like faces and long flowing tuxedos. It is a reflection of the city, in this case Berlin, as a centre for corruption. Raykowski has not only borrowed the title, but his painting composition evokes the bright, bold impasto painting technique of Kirchner, yet without his gestured, expressionistic style. Instead, Raykowski takes Kirchner's original composition and reduces his slabs of colour to built-up blobs of pure colour stacked upon one another like a layered tower.

In art, as is the situation in society, the adage "size matters" holds true. Often artists and audiences have entrenched ideas about art, one of which is to equate "size with quality". Size has become an all-too important criterion for differentiating a monumental masterwork from the plethora, a convention Raykowski attempts to destabilise. Think of the sheer scale of land-based art, Cristo's Wrapped Coast, Jeff Koon's Puppy or the expansive canvasses of Pollock and imagine their importance had they been created in just a hundredth of their scale. Toying with this idea, Raykowski's canvasses measure merely 2.5 centimetres square and his protrusions 14-15 centimetres in length. Deliberately this is also the average length of the male penis, making the work very playful and sexually suggestive, especially considering that the phallic-like protrusions tend to become flaccid and flop over time. Their tendency to "go limp" is due to the suppleness and malleability of the acrylic material, which is yet to harden and solidify. Until such time, Raykowski's protrusions must be periodically checked and cared for by the curator, as he explains:

Each of the three tiny paintings has two holes on the back of the frame. They are intended for hanging the canvas on the wall on one of the three nails provided for this purpose. The holes enable the painting to be mounted upside down when required. It is essential that the painting, which is the most flaccid, be identified twice daily and then turned 180 degrees around. This procedure should be repeated every day preferably at the time of the gallery opening and its closing.[4]

Ironically, the curator's hand is responsible for erecting the artworks and maintaining their erections for the duration of the exhibition. Thus, Raykowski sees his work "not as an object (for example a painting) but as performance, an interactive piece with its own life, rhythm, and demise, hence the instructions to the curator". Evidently, Raykowski is dealing with a variety of disparate social issues and myths. He not only blurs the distinction between painting, sculpture and performance as artistic genres, but he alludes to the perceived male dominance of the painting field throughout history and endeavours to undermine the glamour, power and prestige surrounding the painting genre.

Sven Knudsen

---

[4] Artist's instruction for the curator (displayed to the right of the artworks).

LIST OF FIGURES RELATED TO ARTWORK CREATED BY THE AUTHOR

Fig 1: Raykowski W, 2002, *House at Nobby beach*, sand and seawater, 70 cm x 35 cm x 70 cm; nonextant

Fig 2: Raykowski W, 2002, *One-metre long construction timber*, pine timber, 180 cm x 80 cm x 6cm; nonextant

Fig 3: Raykowski W, 2002, *Untitled*, (Invaginated figures series), Acrylic on canvas, 60 cm x 120 cm

Fig 4: Raykowski W, 2002, *Narcissus,* (Invaginated figures series), Acrylic on canvas, 90 cm x 60 cm

Fig 6: Raykowski W, 2002, *Red and White*, (Invaginated figures series), Acrylic on canvas, 50 cm x 50 cm

Fig 7: Raykowski W, 2002, *Red square*, (Invaginated figures series), Acrylic on plaster board, 75 cm x 88 cm

Fig 8: Raykowski W, 2002, *Standing up*, (Invaginated figures series), Acrylic on plaster board, 55 cm x 120 cm

Fig 9: Raykowski W, 2002, *Black rectangle and a thick line* (Invaginated figures series), Acrylic on canvas, 90 cm x 60 cm

Fig 10: Raykowski W, 2004, *A bagful of air,* (Invaginated figures series), plastic bag, fan, crate and label, 90 cm x 60 cm x 60 cm

Fig 11: Raykowski W, 2005, *Every time you enter the gallery the air of your volume leaves the room*, seven plastic boxes and label, 157 cm x 35 cm x 22 cm

Fig 12: Raykowski W, Show: *AU$29 300*, signage on the front of the gallery

Fig 13: Raykowski W, Show: *AU$29 300*, overall view

Fig 14: Raykowski W, *Painting that doesn't exist no 4*, (Show: *AU$29 300*), medium undetermined, size unknown

Fig 15: Raykowski W, 2002, *Distance from the ceiling to the plinth*, 20 meter-long tape on 30 cm x 30 cm x 80 cm plinth, label, Size: 551 cm

Fig 16: Raykowski W, 2005, *"a little more to the left"*, acrylic on string and canvas, a nail, nail-holes and label; 140 cm x 84 cm

Fig 17: Raykowski W, 2005, *Artist's height as marked on 7 July 2001*, acrylic on canvas, label, 50cm x50cm

Fig 18: Raykowski W, 2004, *Ten feet long tape (footprints of the artist walking over the sticky tape)*, adhesive tape on the wall, 267 cm x 4.8 cm, nonextant

Fig 20a: Raykowski W, 2004, *Oval section 1*, acrylic on canvas, 20cm x 25 cm

Fig 20b: Raykowski W, 2004, *Oval section 2*, acrylic on canvas, 20cm x 25 cm

Fig 20c: Raykowski W, 2004, *Oval section 3*, acrylic on canvas, 20cm x 25 cm

Fig 20d: Raykowski W, 2004, *Oval section 4*, acrylic on canvas, 20cm x 25 cm

Fig 21: Raykowski W, *Society is what becomes of individuals when they get together*, print (Spencer Tunick's work) and label, 45 cm x 52 cm

Fig 22: Raykowski W, 2004; *Two sides of the coin (Separating heads from tails)*, Anxiety series; AU$2 coin split in half, 2 x (2 cm x2 cm)

Fig 23: Raykowski W, 2004; *Erased dollar note, (Erased face of US$1 bill no B90077927 A)*, Anxiety series; US$1 note on board, 15.7 cm x 6.6 cm

Fig 24: Raykowski W, 2004; *Two dollars, (Two One Dollar notes fused together face to back)*; Anxiety series, note no E 38927281 F and B 91870067 D; 15.7 cm x 6.6 cm

Fig 25: Raykowski W, 2004; *A half of a metre, (One metre long measure tape cut in the middle along its entire length)*; Steel measure tape, 1 metre

Fig 26: Raykowski W, 2005; *Untitled, (I hate painting)*; a half-hour-long performance; acrylic on acrylic painted wall, masking tape, video player on plinth;

Fig 27: Raykowski W, 2007, "*13 14 15 22 28 30 32 38 40 – 26 28 37 41 43 44 45 – 3 5 7 8 14 23 31 39 – 2 4 6 8 9 18 27 33 36 41 43 45 – 1 2 11 20 29 36 38 40 41 43 – 34 36 38 40 – 1 3 5 7 9 17 26 35 38 39 – 1 3 6 33 35 37 39 – 33 35 37 39 – 6 13 20 27 33 34*", lotto coupon and label, 18.5 cm x 26 cm

Fig 28: Raykowski W, 2011; *$40*; 3969 ten-cent coins and label; 148 cm x 148 cm

Fig 29: Raykowski, 2005; *Calm Sea*; acrylic on canvas, string and nail; 65 cm x 70 cm

Fig 31: (From left to right) *Berlin Street Scene with Red Streetwalker; M12; Still life with Mortar, Jug and Copper Saucepan*; 2001-2005, acrylic on canvas; 2.5 cm (width) x 2.5 cm (height) x 15 cm (depth)

Fig 33: Raykowski W; 2005; *Four arrows;* acrylic on canvas and string, nails, label; 105 cm x 110 cm

Fig 34: Detail of (Raykowski W; 2005; *Four arrows;* acrylic on canvas and string; 105 cm x 110 cm)

Fig 35: Raykowski W; 2011; *Text;* acrylic on canvas and string, nails and label; 60 cm x 120 cm

Fig 36: Raykowski W; 2005; *Chair;* acrylic on canvas and string, 20 nails and label; 87 cm x 96 cm

Fig 37: Raykowski W; the photograph shown in the appendix B was taken by Wes Raykowski at ***Collab*eratum: *UQCA*@metro** exhibition opening night.

www.ingramcontent.com/pod-product-compliance
Lightning Source LLC
Chambersburg PA
CBHW040746200526
45159CB00023B/1742